*How to Stay Connected in a
Technology-Based World*

Kathleen Barbier

Sweety Mail is a sensational idea for keeping the lines of communication open without hitting information overload. A choice few words can say volumes and Kathy has provided them for us. All we have to do is copy and paste! Keep the sparks flying in just a few keystrokes. The brief messages are sincere and heartfelt without sounding rushed. What a brilliant idea.

Lynn Torre CFP®
Author of *Dance of the Leaf* -
Poetic reflections on life, love, & soul

"Sweety Mail in a Wigged-Out World" not only provides entertainment and a fun game, but also speaks to the importance of keeping relationships fresh and connected in today's high-technology world. Sometimes relationships get lost in the electronic noise and clutter. Kathy gives us a fun and creative way to change that!"

-- Gwendolyn Geer Field
Author, The Butterfly's Kingdom

Sweety Mail (SMs) in a Wigged-Out World

How to Stay Connected in a Technology-Based World

Kathleen Barbier

Table of Contents

Introduction

Today, we are **OVERWHELMED** and **Stressed Out** with too much communication.

There are too many texts, emails, FAXes, phone calls, billboards, TV Ads, and other communication "noise" assaulting us 24/7 in our crazy world.

We simply want it to stop. *Or, do we?*

What we will NEVER HAVE ENOUGH OF is short, loving "just wanted to let you know I was thinking of you" messages from our loved ones.

We need it to keep our relationships fresh and alive, to know that we are BIGGER THAN LIFE in the eyes of our loved ones.

Let me share with you a contemporary Fairy Tale:

Contemporary Fairy Tale

A TWENTY-FIRST CENTURY FAIRY TALE GONE BAD, THEN SAVED --

nce Upon a Time, in a faraway place, Prince Robert met Princess Kathleen on an electronic dating site.

Everything went perfectly with this Match (hint) until they tried to fit their new-found relationship into their everyday living.

ALAS, to their mutual consternation, the communication started to break down as Prince Robert TRIED to communicate his undying love via email.

Princess Kathleen, being the contemporary Princess that she was/is, began to see long (we could define this more precisely, but I think you get the picture, as I see that you are already chuckling) emails on her Smart Phone from the moment she opened her wee Princess eyes in the morning.

DOUBLE ALAS, as even before sipping on her morning Mocha, Princess Kathleen's waking brain began to short circuit. Sizzle. Sputter. ZZzzzz. Crash!

Her love for Prince Robert was being clouded by the information overload which followed.

GULP. The messages came often, with great, loving detail, resulting in a communication overload of Princess Kathleen, who already received 200+ emails, texts, calls, and Faxes in her Wigged-Out World.

Her knee-jerk reaction (not Princess like) was to send a counter reply … "Please, I don't need to know your every waking thought."

Yes, it is sad, but true that the hastily sent email struck her Prince's loving heart squarely in its hopeful, gently beating core, striking a blow against love (OK, they are not young, but their relationship is).

Alas, even a Princess has her "bad hair" days!

REACTION, NOT RESPONSE was the order of the day as Princess Kathleen first declared a moratorium on all emails coming from her loving Prince Robert.

Ah.   What to do?   Prince Robert was sad that he could not express his loving thoughts to his fair Princess.

In this Tale, it was Princess Kathleen who had turned into the Toad. RIB-itt. RIB-itt. (A toad SM)

The Solution – A Game is Born

BUT, YET, the Sun again arises in the Electronic Dating Kingdom as a new game is born.

The repentant Princess Kathleen made a decree upon the land that, from here forward, she and Prince Robert would have a game to play to keep their loving communication alive and well.

This game would be their own, private game, and it would enable them to stay connected in a Wigged-Out World.

THE NEW GAME WAS CALLED "SWEETY MAIL", and the mode of communication was called Sweety Mails (SMs).

The Rules

The game rules were easy … all SMs would have no more than 20 words, but each SM sender got unlimited rights to send SMs to their heart's content throughout the day.

Bugles sounded throughout the land.

Can we see hearts beating stronger again? Do we see certain other forms of communication also being enhanced? Oops. Did I really say that?

AND, AS THE SUN SETS IN THE ELECTRONIC DATING KINGDOM where our fair Princess and her Prince continue their Match unabated, we see Prince Robert and Princess Kathleen happily exchanging snippets of information whenever their loved one comes to mind.

The sun again rises.

Now, doesn't this remind you of being a teenager and being able to pass a note in class without getting caught?

Imagine an SM sneaking into the middle of a Board Meeting?! Such joy.

TODAY, THE PRINCE AND HIS PRINCESS HAPPILY open their SMs the moment they receive them, knowing that it will be a short, entertaining, loving exchange.

The SM will be their "Power Boost" for the day.   It will make them smile.

They know that they are loved and cared about.

In a world that discounts us all, the SM receivers KNOW their true worth.

Princess Kathleen has even declared it a better way to start her day than her traditional Mocha!!

HARMONY ONCE AGAIN CAME OVER THEIR KINGDOM.

And, Prince Robert and Princess Kath-leen live happily ever after, texting 2.5 SMs per day, on average, to a new "EX-PERT SKILL LEVEL" of 10 words or less!

We see a rainbow across the sky as Prince Robert leans forward to kiss his blushing Princess.

❤ The End

Purpose

The purpose of this book is to inspire loving interchanges in relationships, whether those relationships are brand new or the couple has been together long enough to have grown together in magical ways.

We understand that electronic dating sites are giving the book freely to their matches, hoping that the new, enhanced communication will increase their client satisfaction ratings among their matches.

SM has energized emails, and replaced "texts" as the preferred couple talk tool of the twenty-first century!

And, in the words of Princess Kathleen …

Let the SMing begin!

May the Force be with you, and your SMs keep peace in your Kingdom!

♥Princess Kathleen

Sweety Mail (SMs) for the Novice

GETTING STARTED

Directions:

The purpose of this book is to get you started in the Sweety Mail game with minimal start up time.

Therefore, this book is filled with a "starter set" of SMs, which are 20 words or less in length.

Where do I start?

If you're not quite sure how to start, look through the Table of Contents, and find the category that speaks to you.

Follow your heart, and you will know where to begin.

Before long, you will be SMing with the best of them!

Also included are "Expert Level" examples, using 10 words or less, when time is of the essence!

STEP 1 –

SEND THE FIRST SWEETY MAIL

To get started: Type "Sweety Mail (SM) Program" in Subject Line.

Subject: Sweety Mail (SM) Program:

(Note: Text for SM program explanation would follow)

Hi, ____ (name), I've got a new idea to make our personal communications more fun.

I understand that it is intended to be a game, and in a second, hopefully you will see why.

I think we can have a lot of fun with it, and when we receive a Sweety Mail (SM), we will know that it will be a short, but fun break in the day!

How it Works

We each get unlimited SM opportunities in a day, but each email cannot exceed 20 words. I think it will be fun!

Want to give it a try?

Here's an example:

New fun idea … a game. We each get unlimited SM opportunities in a day, but each cannot exceed 20 words.

Words: 20

℘

STEP 2 –

IMMEDIATELY FOLLOW WITH ANOTHER SM.

HINT: LEAD BY EXAMPLE.

In this example, conflict is growing over misunderstandings as one person is able to send long emails, but the other doesn't have the time to fully read and understand them!

SM #1 (Conflict)

Subject: My SM to You

Darling, don't be so hard on yourself. We are quite different, and that isn't bad! Love you.

Words: 17

℘

Sample Sweety Mail (SMs)

The following pages are filled with examples of SMs that can be used as is, modified, or used to inspire.

Be creative. Be sincere. But, most importantly, be succinct!

Remember, the 20 words you choose to send will shape your emotional well-being today, tomorrow, and in the future.

The 2 minutes you take to send a SM will add "money in the bank" with your sweetheart.

It is the wisest investment of your time that you can possibly make.

When you consider that relationships can endure, and even thrive for 30 or 40 years, imagine what is possible!

A SM of apology will be worth its weight in gold.

A flirty SM sets the stage for a glorious evening.

It is all up to you. What will it be?

To Deal with Conflict

SM #2 (Early Morning Conflict)

Today was clouded by my memory of the misunderstanding we started the day with.

Forgive me?

Words: 16

SM #3 (Conflict)

I can't remember what I did this morning, but apologize anyway.

Clearly, it was not as important as YOU are!

Words: 20

SM #4 (Verbal Conflict)

I know that I said something stupid this morning, but am not clever enough to even remember what it was!

Words: 20

❦

SM #5 (Conflict)

EXPERT LEVEL

Sorry. Can you forgive me?

Words: 5

❦

SM #6 (Conflict Recovery)

EXPERT LEVEL

Dinner out or dog house in ??

Words: 6

SM #7 (Conflict Recovery)

EXPERT LEVEL

Flowers or grovel ??

Words: 3

To Start the Day Off Right

SM #8 (Morning Thought)

Thinking of you this morning. Just wanted to let you know.

Words: 11

❦

SM #9 (Morning Pause)

Stopped for a moment just to think about us.   Made me smile!

Words: 12

❦

SM #10 (Morning - Tonight)

This morning is filled with thoughts of seeing you tonight. Can't wait!

Words: 12

SM #11 (Morning - Lunch)

The best thing about morning is that it gets me to lunch with you!

Looking forward to our interlude!

Words: 19

SM #12 (Morning break)

EXPERT LEVEL

Thinking of you as my morning break.

Word Count: 7

That Important "Morning After" SM

SM #13 (Morning After)

Last night added new meaning to "seeing stars." Can't wait to see you again.

Words: 14

℺

SM#14 (Morning After)

Unscheduled surgery today to remove smile from my face and reconnect brain!

Words: 12

℺

SM #15 (Morning After – Blue eyed)

Blue like the color of your eyes is how I feel this morning without you.

Words: 15

❧

SM #16 (Morning After – 4th of July)

Loved celebrating the 4th of July early with our special fireworks last night!

Words: 13

❧

SM #17 (Morning After)

EXPERT LEVEL

Thanks for a spectacular night. Be thinking of you today.

Words: 10

SM #18 (Morning After)

EXPERT LEVEL

You've left me speechless.

Words: 4

SM #19 (Morning After)

EXPERT LEVEL

Bravo! Repeat performance ??

Words: 3

❦

SM #20 (Morning After)

EXPERT LEVEL

Unprecedented and unparalleled !!

Words: 3

❦

SM #21 (Morning After)

EXPERT LEVEL

Unbelievable !!

Words: 1

SM #22 (Morning After)

EXPERT LEVEL

Your place or mine?

Words: 4

When You Have to Cancel Plans

SM #23 (Cancel Breakfast)

Devastated that I won't be able to start the day over breakfast with YOU!

Reschedule?

Words: 15

SM #24 (Cancel Breakfast)

Hate to start the day without seeing you, but got called in early.

Forgive me?

Words: 15

SM #25 (Cancel Lunch)

Just got major project dumped on my desk that wiped out lunch. Can you forgive me? Rather be with you.

Words: 20

SM #26 (Cancel Lunch)

How's a guy supposed to function without seeing his girl at lunch !? Called into meeting. Dinner tonight?

Words: 17

SM #27 (Cancel Dinner)

Not fair! Pulled into meeting. Have to give up dinner with my best girl !

Call when I escape!

Words: 18

SM #28 (Cancel Multiple Dinners)

If I have to miss one more dinner with you, I'll have to take you on a 3-day weekend instead !!

Words: 20

SM #29 (Cancel Dinner - Reschedule)

EXPERT LEVEL

Sorry.   Reschedule to favorite restaurant next week?

Words: 7

SM #30 (Cancel Dinner - Paris)

EXPERT LEVEL

Reschedule to Paris, s'il vous plait?

Words: 6

Romantic SMs

SM #31 (Romantic Rendezvous)

Near your place this afternoon. The idea of seeing you mid-day rocks me. Time for a quick rendezvous?

Words: 18

SM #32 (Romantic)

You captured my heart, enabling me to declare the 3 most beautiful words in life – I love you!

Words: 18

SM #33 (to a blonde)

Repunzel, Repunzel, let down your golden hair. Must see you!

Words: 10

SM #34 (Wishes)

Your love opened my life and heart so completely that my saved-up wishes are rapidly coming true.

Words: 17

SM #35 (Romantic Footsteps)

The racing pitter-patter of childhood foot-steps is so like the pitter-patter of my heart whenever I think of YOU!

Words: 19

❧

SM #36 (Romantic days and nights)

A thousand days and a thousand nights were not too long to wait for you. It only seemed forever!

Words: 19

❧

SM #37 (Love Waiting)

I've waited a long time for you to steal my heart! I am comforted knowing you will return it unharmed!

Words: 20

SM #38 (Love Songs)

When I look into your eyes, my heart begins singing love songs. Will be ready for American Idol soon!

Words: 19

SM #39 (Dreams come true)

With every dream you've made come true, my love grows stronger.

Words: 11

SM #40 (Snow White)

Even without looking into your mirror, I can tell you that you are, indeed, the fairest maiden in the land!

Words: 20

SM#41: (Sleepless in Seattle)

Our date reminded me of Sleepless in Seattle, like I was coming home again. Looking forward to seeing you soon.

Words: 20

SM #42 (Tongue Tied)

My appreciation for you is exceeded only by my inability to fully express it!

Words: 14

SM #43 (to a brunette)

EXPERT LEVEL

Snow White, abandon the dwarfs.   Meet me at 8!

Words: 9

No Time to Call

SM #44 (No time to talk)

Would call just to hear your lovely voice but my cell battery is dying. Kisses.

Words: 15

SM #45 (No time to talk)

Would call just to hear your sexy voice but then I'd be distracted all day! Long, slow kisses.

Words: 18

SM #46 (No time to talk)

Would call if I had even one free second, but will have to send a kiss instead!

Words: 17

Just Because I was Thinking of You

SM #47 (Just because)

Keep clicking heels together, repeating "there's no place like home." Must have wrong shoes on today.

Words: 16

℘

SM #48 (Just Because)

EXPERT LEVEL

Just wanted you to know that you're on my mind!

Words: 10

℘

SM #49 (Just Because)

EXPERT LEVEL

Just thinking of you, Sweetheart!

Words: 5

℘

SM #50 (Just Because)

EXPERT LEVEL

Kisses to my one in a million!

Words: 7

℘

SM #51 (Just Because)

EXPERT LEVEL

Be still my heart!

Words: 4

Time to Celebrate

SM #52 (Celebrate Salary Bonus)

Would trade bonus I just got for 5 minutes with you. Meet me at our favorite place to celebrate.

Words: 19

SM #53 (Surprise Celebration)

Too excited to spoil the surprise. Meet me at our favorite place !!

Words: 12

SM #54 (Month Anniversary)

Time to celebrate your coming into my life. Can't believe it has been one month already!

Words: 16

Ｑ

SM #55 (1 Year Celebration)

Hard to believe my great fortune as you took my breath away a year ago. Meet to celebrate?

Words: 18

Ｑ

SM #56 (6 Month Anniversary)

Six months of joy and happiness. Let's go away this weekend to celebrate!

Words: 13

❦

SM #57 (1 Year Anniversary)

One year of joy and happiness. Let's steal away early from work today to celebrate!

Words: 15

❦

SM #58 (2+ Year Anniversary)

Darling: Thank you for the best ___ (insert #) years of my life. Can't wait for our celebration dinner tonight!

Words: 18

℘

Asking for Help

SM#59 (Laundry)

Would be lost without you. Leaving early, and don't have a prayer of picking up laundry. Can you save me?

Words: 20

❦

SM#60 (Doggie Sitting)

Rover asked if you would visit him while I'm gone. He promised to keep an eye on you for me!

Words: 20

❦

SM # 61 (Forgotten Kids' Weekend)

Forgot it was my weekend to play Dad. Hoping you can join us for a day in the park.

Words: 19

℘

When Feeling Rebellious

SM #62 (Rebellious)

Iwasthinkingofyouthismorningandknewth-atIjustcouldn'texpresshowmuchyoumean-tomeintwentywordsorlesssoIjustecidedto-doitinonewordaccordingtotheSMRules.

Words: 1

SM #63 (Feisty)

I know you said something really nice to me this morning, but I think I heard it differently, Princess.

Words: 19

SM #64 (Feisty)

Know you thanked me for all the work I did this weekend, but I can't remember hearing it.

Words: 18

SM #65 (Feisty & Memory Problem)

If I could remember what I was angry about this morning, I would apologize. But, how about "LOVE YOU" instead?

Words: 20

Checking to See if All is Well

SM #66 (a Pre-call check after a fight)

EXPERT LEVEL

Having a good day?

Words: 4

❦

SM #67 (Home late pre-call check)

EXPERT LEVEL

Has that beautiful smile returned?

Words: 5

❦

SM #68 (Making up Roses in Order)

EXPERT LEVEL

Trade red roses for badly-needed hug.

Words: 6

SM #69 (Something Stupid)

EXPERT LEVEL

Was that a nightmare, or me acting out last night?

Words: 10

SM #70 (Forgiveness)

EXPERT LEVEL

Pleeeeeease forgive me (again)?!

Words: 4

When You're Running Late

SM #71 (Running Late)

EXPERT LEVEL

Runny late, Fair Maiden, but please don't escape !!

Words: 8

SM #72 (Late)

EXPERT LEVEL

I'm late. I'm late for a very important date!

Words: 9

SM #73 (Late)

EXPERT LEVEL

Sorry. No words can express ... Be there soon!

Words: 8

❧

SM #74 (Late)

EXPERT LEVEL

Princes are NEVER supposed to be late, fair Princess! Sorry!!

Words: 10

❧

SM #75 (Late)

EXPERT LEVEL

There are no excuses for making you wait, just regrets.

Words: 10

In summary, it is important to pay attention to the relationship between the seriousness of the issue, and the shortness of the SM.

The more serious the issue, the more important it is to keep the message short!

More words increase the odds of a misunderstanding when someone is overly stressed!

Good luck and Good Love! Princess Kathleen

KATHLEEN BARBIER parlayed a successful business career as an executive with Fiserv, Inc. responsible for Marketing, Sales, Client Services, and Corporate Communication, into the establishment of Estate Doctor®, a wealth management business focused on helping women who have undergone sudden loss, and who must for the first time in their lives learn how to manage their financial affairs.

Born in Carmel, California, Kathleen earned advanced business degrees (MBA and a Master of Science – Finance), taught graduate and undergraduate School of Business courses, and became a skilled change agent in the public, private, and non-profit sectors.

Kathleen believes that communication is the key to much of life's challenges. While working in the public sector, Kathleen had the opportunity to work with hundreds of city employees to help them embrace change and let go of the things holding them back.

Kathleen lives in Pebble Beach, California and has two grown children who are married.